I0502538

6th Grade Math
Volume 4

© 2013 OnBoard Academics, Inc
Newburyport, MA 01950
800-596-3175
www.onboardacademics.com

ISBN: 978-1494857318

ALL RIGHTS RESERVED. This book contains material protected under International and Federal Copyright Laws and Treaties. Any unauthorized reprint or use of this material is prohibited. No part of this book may be reproduced or transmitted in any form or by any means, electronic or mechanical, including photocopying, recording, or by any information storage and retrieval system without express written permission from the author / publisher. The author grants teacher the right to print copies for their students. This is limited to students that the teacher teachers directly. This permission to print is strictly limited and under no circumstances can copies may be made for use by other teachers, parents or persons who are not students of the book's owner.

Table of Contents

Finding the Percent of a Number

Key Vocabulary

percent

interest

Match these "easy" percents.

1%		$\frac{1}{4}$
10%		$\frac{1}{2}$
50%		$\frac{1}{10}$
25%		$\frac{1}{100}$

Use the ———————— **"easy" percents to complete this graph.**

Table A

50% of 360	360 ÷		=	
25% of 360	360 ÷		=	
10% of 360	360 ÷		=	
1% of 360	360 ÷		=	

Use your answers from Table A to complete Table B.

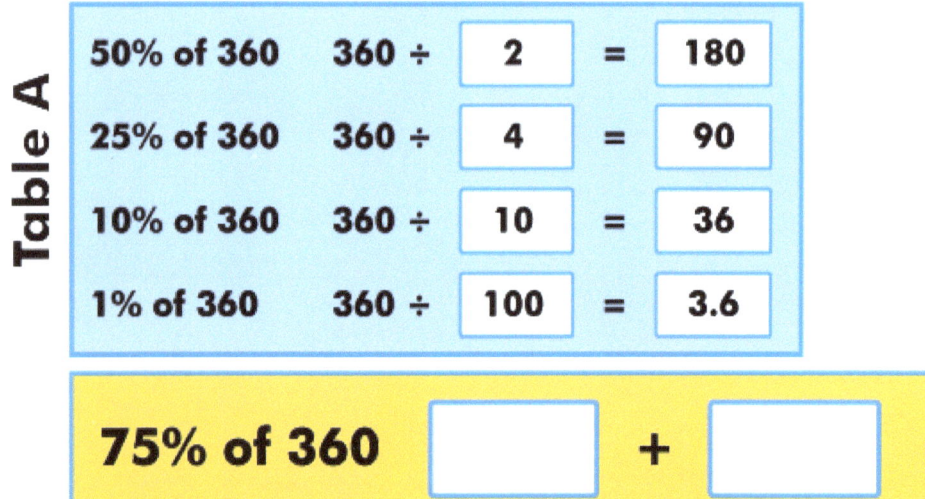

Table A

50% of 360	360 ÷	2	=	180
25% of 360	360 ÷	4	=	90
10% of 360	360 ÷	10	=	36
1% of 360	360 ÷	100	=	3.6

Table B

75% of 360		+		=
5% of 360		÷		=
11% of 360		+		=
9% of 360		–		=

Buying a New Computer

Ben purchases his laptop on the payment plan at Laptops-U-Like. Ashima purchases the same model, also on the payment plan, at Loads-of-Laptops.

How much interest do they each pay in total?

Finding a Percent of a Number Using Decimals.

Study the illustration below.

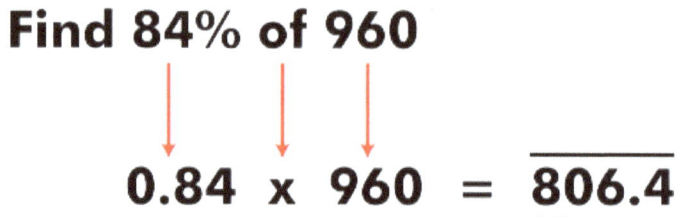

Find 84% of 960

$$0.84 \times 960 = \underline{806.4}$$

Write the percent as a decimal and multiply by the number.

Find 7% of 600

Name_____
Finding the Percent of a Number Quiz

1 True or false? An easy way to find 10% of a number is to divide the number by 10.

2 75% of 180 is:

A 135

B 140

C 145

D 155

3 What is 1% of 1,080?

4 What is 16% of 1,080?

Proportion

Key Vocabulary

proportion

equivalent proportion

cross products

Chocolate Smoothie Recipe

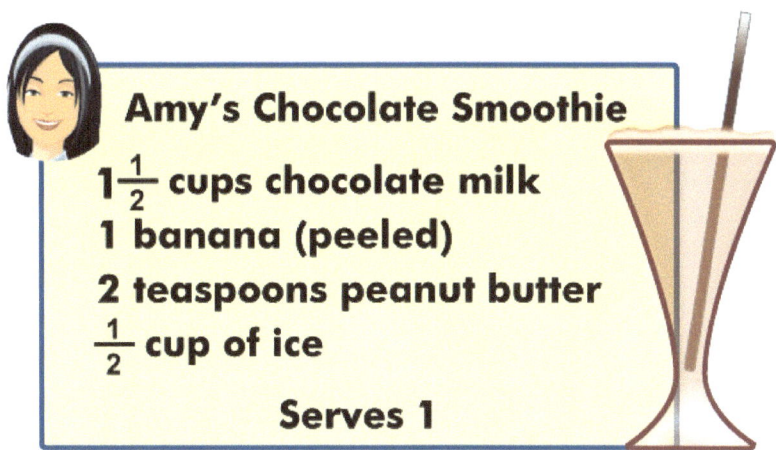

Amy is making Chocolate Smoothie for herself and two friends. How many teaspoons of peanut butter will she need?

_____

A table can help solve proportion problems.

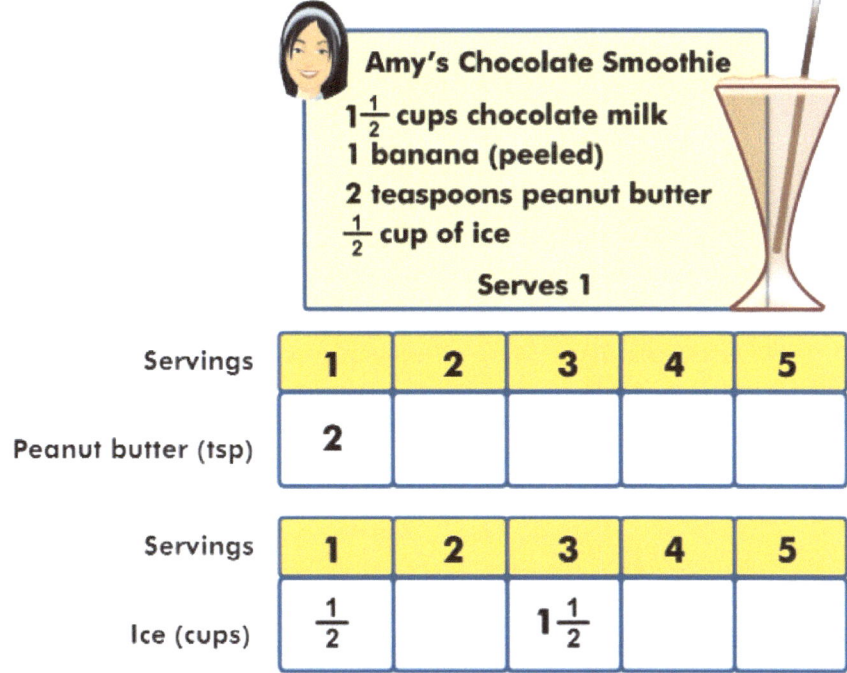

Servings	1	2	3	4	5
Peanut butter (tsp)	2				

Servings	1	2	3	4	5
Ice (cups)	$\frac{1}{2}$		$1\frac{1}{2}$		

Calculate the correct amount of chocolate and yogurt

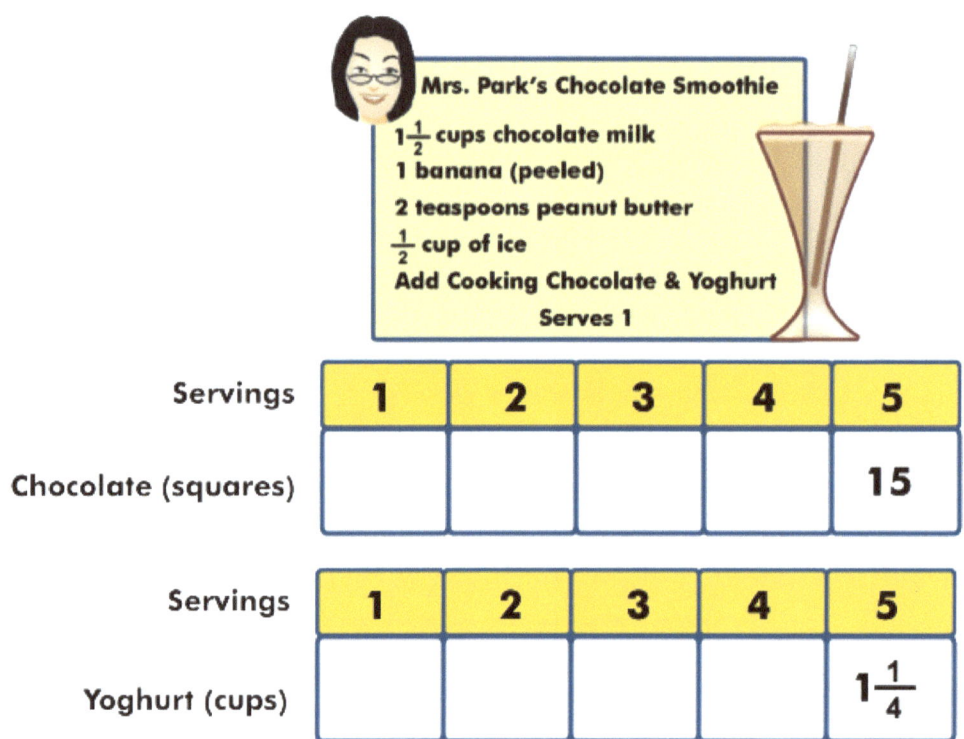

Mrs. Park's Chocolate Smoothie

$1\frac{1}{2}$ cups chocolate milk
1 banana (peeled)
2 teaspoons peanut butter
$\frac{1}{2}$ cup of ice
Add Cooking Chocolate & Yoghurt
Serves 1

Servings	1	2	3	4	5
Chocolate (squares)					15

Servings	1	2	3	4	5
Yoghurt (cups)					$1\frac{1}{4}$

You can also solve proportion problems using equivalent fractions.

Servings	1	2	3	4	5
Chocolate (squares)	3	6	9	12	15

$$\frac{1}{3} \qquad \frac{2}{6} \qquad \frac{3}{9} \qquad \frac{4}{12} \qquad \frac{5}{15}$$

What is the real-world distance between El Paso and Dallas?

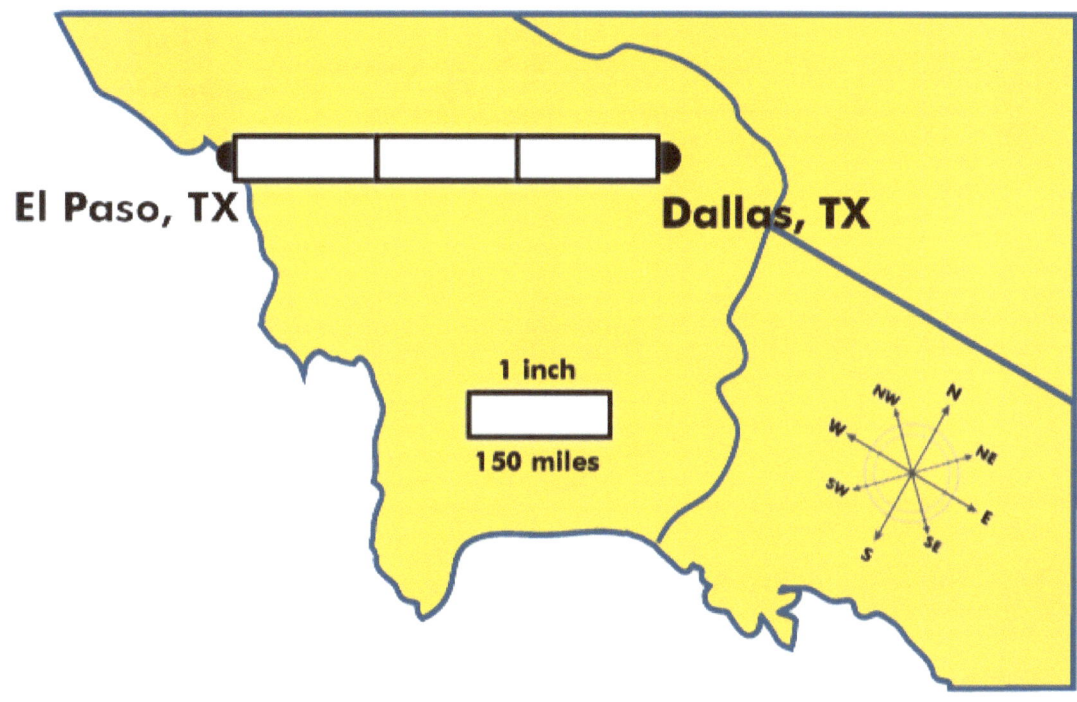

El Paso, TX

Dallas, TX

1 inch

150 miles

Distance on the map

1 inch		1 inch		1 inch		
	+		+		=	
150 miles		150 miles		150 miles		

Real-world distance

1 inch		1 inch		1 inch		
	+		+		=	
150 miles		150 miles		150 miles		

The distance between El Paso, TX and Little Rock, AR is 750 miles.

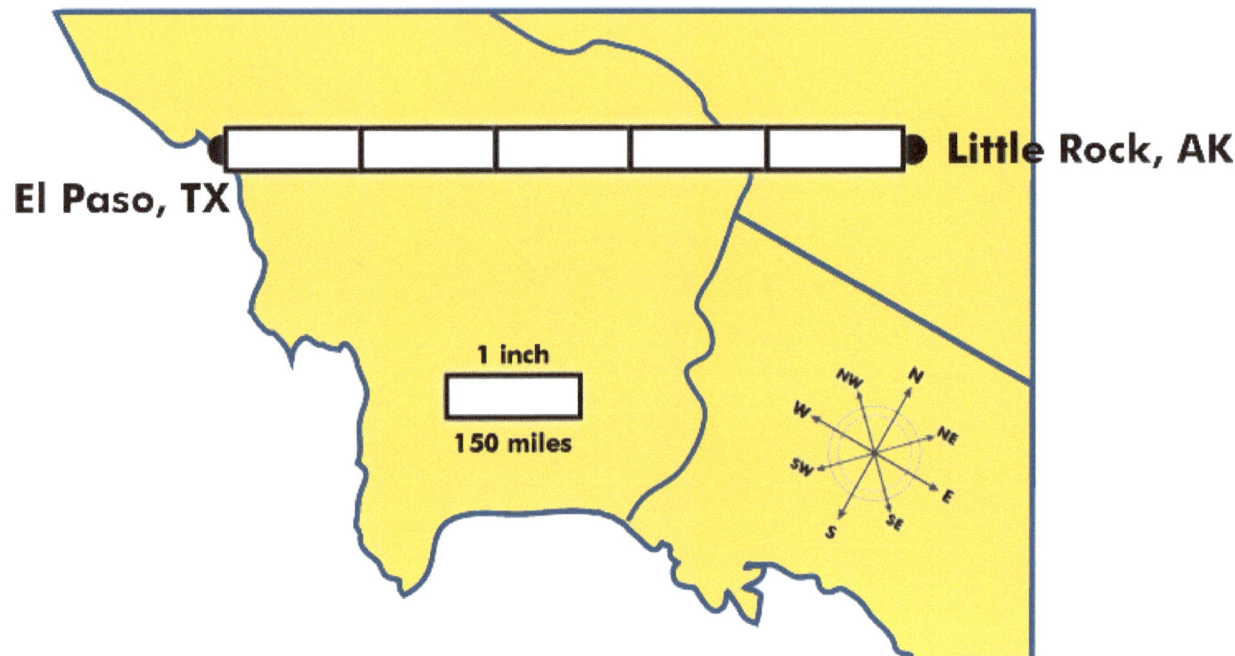

El Paso, TX

Little Rock, AK

1 inch

150 miles

How many inches would this be on the map?

Stretch your knowledge.

Filling Up with Gas

 Jack's Dad stops at the gas station and tops up his tank with 5 gallons of gas. It costs him a total of $19.25. His tank holds 16 gallons.

How much would a full tank of gas cost?

Name_____

Proportion Quiz

1 True or false? $\dfrac{12}{18} = \dfrac{6}{9} = \dfrac{1}{3}$

2 One batch of barbecue sauce requires $\dfrac{3}{4}$ cup of sugar. How many cups of sugar are needed for 5 batches?

- Ⓐ $3\dfrac{1}{2}$
- Ⓑ $3\dfrac{3}{4}$
- Ⓒ 4
- Ⓓ $4\dfrac{1}{4}$

3 Jack paid $3.20 to download 4 songs. What was the cost of each song? Answer in dollars.

4 If 5 bottles of soda cost $2.70, how much do 2 bottles cost? Answer in dollars.

Algebraic Expressions

Key Vocabulary

expression

evaluate

variable

Examples of Algebraic Expressions.

Study the illustration below to discover how to write algebraic expressions.

(1) One less than a number y

$$y - 1$$

(2) Sum of four and a number b

$$b + 4$$

(3) A number y divided by 40

$$\frac{y}{40}$$

(4) A number x decreased by 4

$$x - 4$$

(5) n students organized into seven equal teams

$$\frac{n}{7}$$

(6) Twice a number n

$$2n$$

(7) Six more than three times a number m

$$3m + 6$$

(8) Eight increased by five times a number n

$$8 + 5n$$

Expressions

Michael weighs w pounds. Express, in terms of w, the weight of his three friends.

Sam weighs 5 lb more than Michael.

Sam's weight:

Dan weighs twice as much as Michael.

Dan's weight:

Lee weighs 2 lb less than Michael.

Lee's weight:

Matching Words and Symbols

Write the words in the yellow boxes under the appropriate symbol.

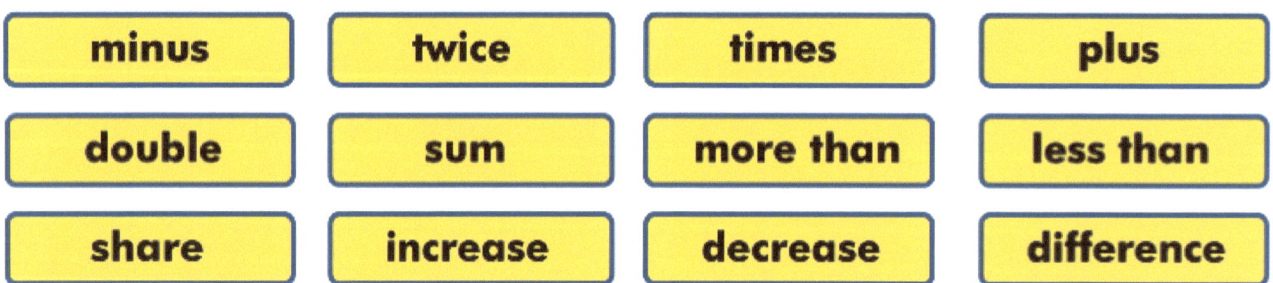

Write these algebraic expressions.

(1) One less than a number y

(2) Sum of four and a number b

(3) A number y divided by 40

(4) A number x decreased by 4

(5) n students organized into seven equal teams

(6) Twice a number n

(7) Six more than three times a number m

(8) Eight increased by five times a number n

Ten Pin Bowling

Write an expression, in terms of *x*, for the number of pins each player has left standing.

Amy knocks down *x* pins.

KJ knocks down twice as many pins as Amy.

Nancy knocks down four more pins than Amy.

Evaluate these expressions.

 Amy knocked over 3 pins (*x*). How many pins does each each player have left standing?

Amy   10 − x

pins
Evaluate

KJ   10 − 2x

pins
Evaluate

Nancy   10 − (x + 4)

pins
Evaluate

_____

a = **2**   b = **3**

2a =

4b =

4b − 2a =

2a − 4b =

Name_____

Algebraic Expressions Quiz

1 True or false: three times a number (n) can be written as n^3?

2 Four more than three times a number (x) can be written as:

A $3(x + 4)$

B $4(3x)$

C $4x + 3$

D $3x + 4$

3 $a = 3$ $b = 4$ $c = -2$ Evaluate $4b - 2a$

4 $a = 3$ $b = 4$ $c = -2$ Evaluate $c^2 - a$

Graphing Linear Functions

Key Vocabulary

linear function

linear equations

graph

table

rule

Squares-complete the table and find the rule

P =

Length of side (s)	Perimeter (p)
1	4
2	
3	12
4	
5	

Complete the graphing of the linear function.

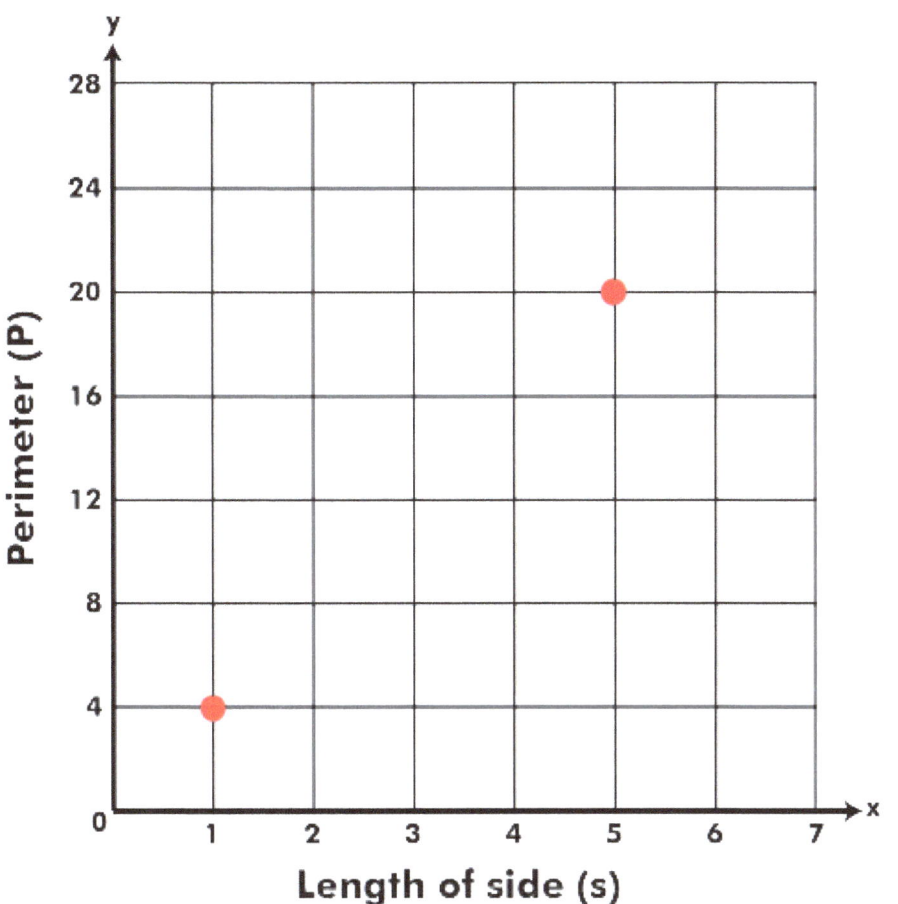

s	P
1	4
2	8
3	12
4	16
5	20

Use the graph to find the perimeter of a square with the length of side 3.5.

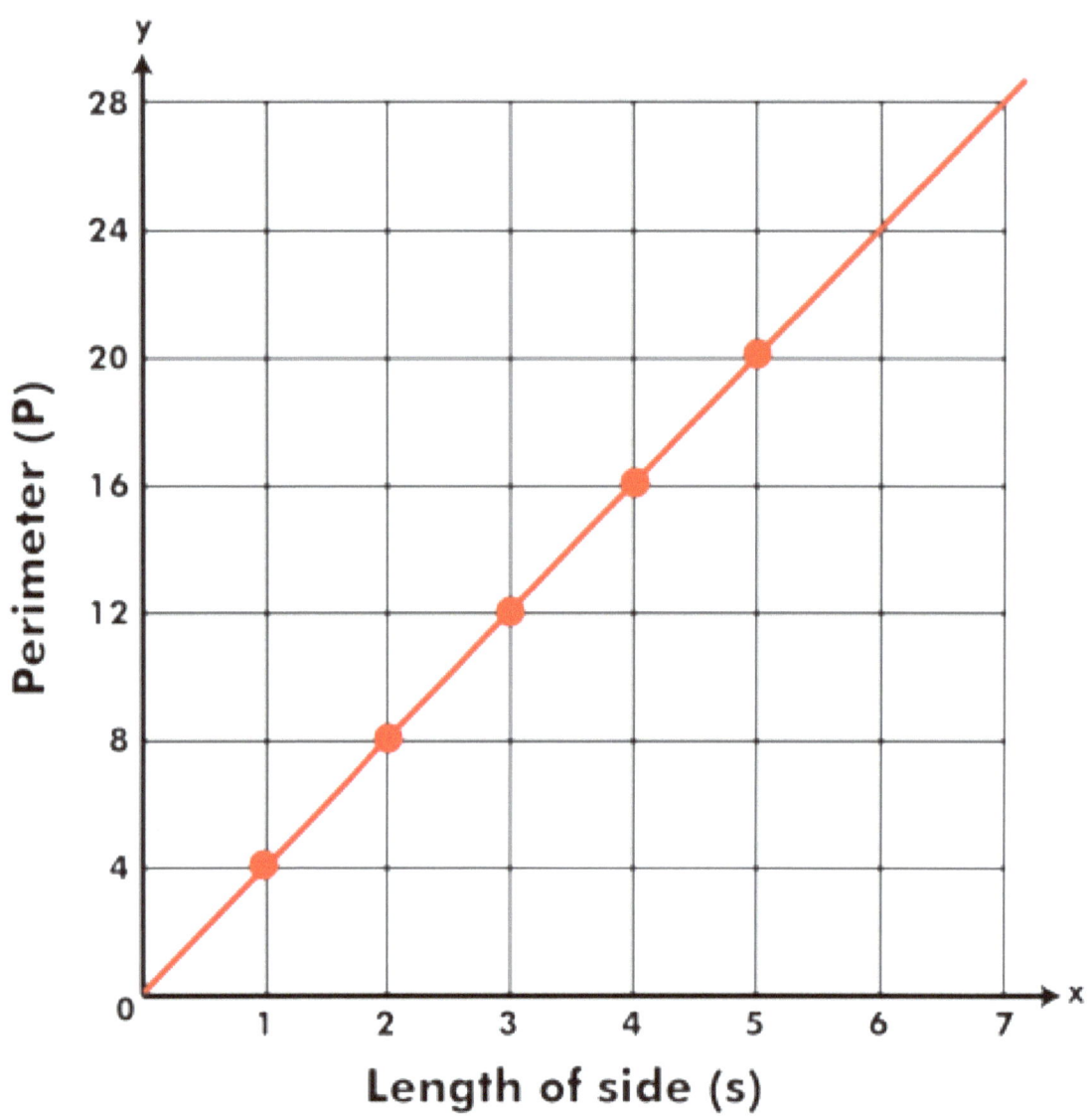

Complete the function tables and find the rules.

x	y
1	
2	5
3	6
4	
5	8

p	m
-4	-6
-2	
0	
2	0
4	

a	b
-3	-6
-2	
-1	-2
0	0
1	

y =   m =   b =

Stretch your knowledge.

A car travels at an average speed of 30 mph

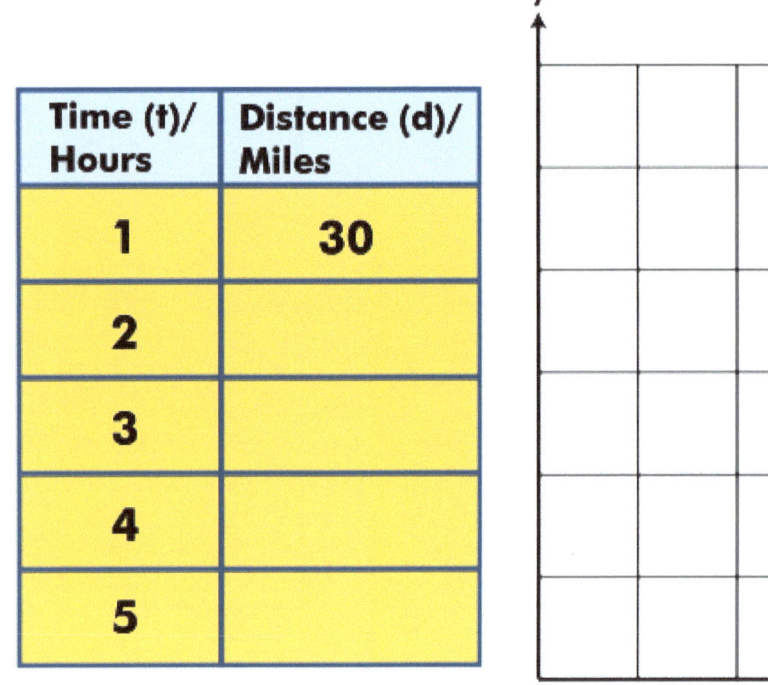

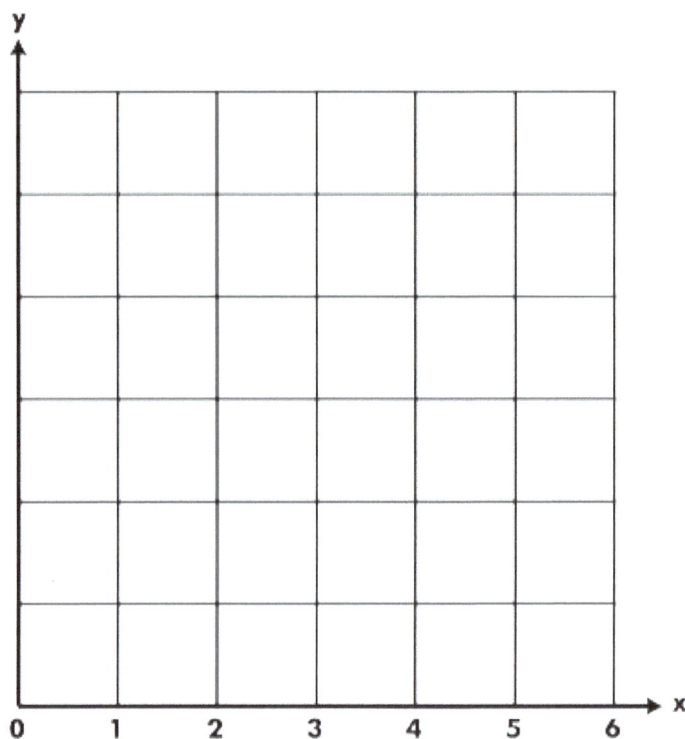

Time (t)/ Hours	Distance (d)/ Miles
1	30
2	
3	
4	
5	

Complete the table, label the grid, graph the function, and find the rule.

Name_____

Graphing Linear Functions Quiz

1 True of false: one dollar is equivalent to 30 rubles?

2 How many dollars do I need to obtain 135 rubles?

- **A** 3
- **B** 3.5
- **C** 4
- **D** 4.5

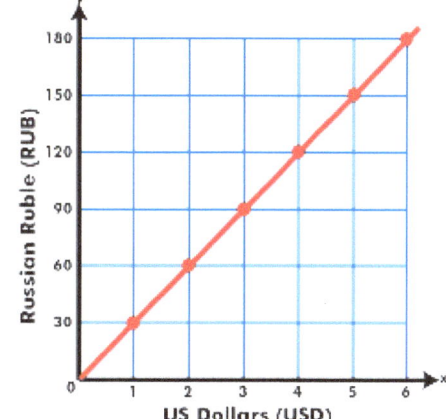

3 What is the missing value in Table 1?

4 What is the missing value in Table 2?

x	y
1	4
2	
3	12

Table 1

x	y
0	
7	4
11	8

Table 2

Newburyport, MA 01950

1-800-596-3175

OnBoard Academics employs teachers to make lessons for teachers! We create and publish a wide range of aligned lessons in math, science and ELA for use on most EdTech devices including whiteboard, tablets, computers and pdfs for printing.

All of our lessons are aligned to the common core, the Next Generation Science Standards and all state standards.

If you like our products please visit our website for information on individual lessons, teachers licenses, building licenses, district licenses and subscriptions.

Thank you for using OnBoard Academic products.